Samy Barros Souza Ibrahim

Sustainability through biogas

Samy Barros Souza Ibrahim

Sustainability through biogas

Biogas via sugar cane and chicken droppings

ScienciaScripts

Imprint
Any brand names and product names mentioned in this book are subject to trademark, brand or patent protection and are trademarks or registered trademarks of their respective holders. The use of brand names, product names, common names, trade names, product descriptions etc. even without a particular marking in this work is in no way to be construed to mean that such names may be regarded as unrestricted in respect of trademark and brand protection legislation and could thus be used by anyone.

Cover image: www.ingimage.com

This book is a translation from the original published under ISBN 978-613-9-72526-7.

Publisher:
Sciencia Scripts
is a trademark of
Dodo Books Indian Ocean Ltd. and OmniScriptum S.R.L publishing group

120 High Road, East Finchley, London, N2 9ED, United Kingdom
Str. Armeneasca 28/1, office 1, Chisinau MD-2012, Republic of Moldova, Europe
Printed at: see last page
ISBN: 978-620-7-92338-0

To my mum (in memorian)

ACKNOWLEDGEMENTS

To the creator;

Special thanks to my wife, partner and friend, Kelley Contieri Silveira Ibrahim.

To my children, Pedro and Mariana Ibrahim, my goals in the struggle;

To my parents, brother and other in-laws, father-in-law, mother-in-law and sister-in-law;

To my confreres, fun guaranteed;

To the students Thalles and Juliana, future teachers;

To my "advisors", Williane and Rodrigo. Without you it wouldn't have been possible;

To my supervisors and co-supervisors, essentially masters;

To my fellow Master's students. They were part of this journey.

My colleagues and friends, Fernanda and Jessica. They allowed everything to happen.

To those who believed in me, thank you.

Everyone has the right to an ecologically balanced environment, an asset of common use to the people and essential to a healthy quality of life, imposing on the public authorities and the community the duty to defend and preserve it for present and future generations.

Article 225 of the Federal Constitution

SUMMARY

The search for renewable energy sources has established sugar cane as an important energy matrix. The generation of bagasse from this input reveals a waste product with high potential for sustainable energy production. With the aim of making better use of this product, this research evaluated the anaerobic production of biogas generated by sugar cane bagasse when chicken droppings were used as an inoculum in experimental batch anaerobic reactors. Six reactors were used, divided into two treatments. The first consisted of *raw* sugar cane bagasse (03 reactors) and the second consisted of sugar cane bagasse treated with NaOH (03 reactors). Chicken droppings were used as inoculum in both treatments. The time taken to read the gases generated was 54 days. It was observed that the point of greatest methanogenic activity for the energy potential of sugar cane bagasse was approximately the twenty-fifth and twenty-sixth day, respectively, for each treatment, and that treatment 2 obtained the best result, producing twice as much methane gas CH_4 in practically the same amount of time.

Key words: Biogas. Sugar cane bagasse. Sustainable energy.

SUMMARY

CHAPTER 1

INTRODUCTION

The search for energy sources that can replace fossil fuels has brought Brazil a unique opportunity. As a potential producer of raw materials, the country has been looking for alternatives in the utilisation of agro-industrial waste in order to adopt sustainable and renewable energy policies.

The aim of replacing fossil fuels is not only to find an alternative source, but also to produce environmentally friendly energy. In this scenario, sugar cane bagasse has emerged.

According to IBGE (2015), Brazil produced 750,107,378 tonnes of sugar cane in 2015 alone. Despite estimating a slight drop in its cultivation figures, around 2.6%, the agency still estimates a productivity of 730,919,055 tonnes for 2016.

According to the Brazilian Sugarcane Industry Association (UNICA, 2015), Brazil planted 10,870,647 hectares of this crop in 2015 alone. Although the high concentration of planting is observed in the centre-south region, the northeast holds 1,248,932 hectares in this chain.

The forecast for sugar cane production in the northeast is 53 million tonnes. Of this total, Pernambuco will account for 13 million tonnes and Alagoas will average 18.5 million tonnes (SINDAÇÚCAR, 2016).

By analysing these figures, you can see that this large cultivated area possibly generates a large amount of waste as well.

In this context, Pelizer (2007) emphasises the importance of appropriate disposal of agro-industrial waste. He emphasises that as well as creating serious environmental problems, waste represents a loss of raw materials and energy.

Different biotechnological processes are being developed to utilise this waste in industrial production, generating products of great economic value. The strategy consists of transforming raw materials into useful goods without damaging the environment, using by-products and emissions as inputs for other products (ISRAEL, 2005).

Sugarcane bagasse, a large-volume waste product of the sugar and alcohol agro-industry, is a source of interest due to its quantity and energy value. Mills and distilleries make use of this input, converting bagasse into energy and steam, burning it in boilers and meeting part of their energy demands (KWhs).

According to Van Haandel & Lettinga (1994), studies carried out on sugarcane bagasse have shown that it is possible to achieve a chemical to electrical conversion efficiency of around 33 per cent (internal combustion reactors) and 50 per cent (turbines).

Another way of utilising this waste is through anaerobic digestion. In this process, the bagasse's organic material is converted into biogas.

The anaerobic process has undergone changes in recent years, but studies aimed at understanding the appropriate quantities, substrates and inoculums to best achieve biogas efficiency and quality have been the subject of several studies.

This research aimed to determine the point of greatest methanogenic activity of sugar cane bagasse, treated and untreated with sodium hydroxide, using chicken droppings as an inoculum.

CHAPTER 2

LITERATURE REVIEW

2.1 Growing sugar cane

Brazil has been growing sugar cane since the time of its discovery, and the first sugar cane plantations were planted with seedlings brought from other continents by the colonisers. Although there is disagreement as to its geographical origin, there is a general consensus among historians that sugar cane comes from South-West Asia, Java, New Guinea and India (FAHL etal., 1998).

According to Lima (1984), the first species introduced to Brazil was *Saccharum officinarum* L., which was brought from the island of Madeira in 1502. This species was known as noble cane or tropical cane, characterised by its high sugar content, large size, thick stalks and low fibre content. Due to these characteristics, S. *officinarum* was cultivated in the first three centuries of colonisation, probably as a single variety, which in the 19th century was given the name "Creoula" or "Mirim" cane, or even "Cana da terra", to distinguish it from the new imported cultivars that began to arrive in the country.

Currently, national policy for sugar cane production is geared towards the sustainable expansion of the crop, based on economic, environmental and social criteria. The Agroecological Zoning of Sugarcane (ZAECana) programme regulates sugarcane planting, taking into account the environment and the economic suitability of the region. Based on a detailed study, areas favourable for planting are stipulated on the basis of climate, soil, biomes and irrigation needs (MAPA, 2016).

According to the Brazilian Sugarcane Industry Association (ÚNICA, 2015), it is estimated that 666,824 tonnes of sugarcane will be crushed for the 2015/2016 Brazilian harvest. Of this total, 16,382 tonnes will be produced by the state of Alagoas (Table 01).

Table 01 - Sugarcane crushing - 2015/2016 harvest

	Highlight	*Sugarcane (thousand tonnes)*
Centre-South Region	São Paulo	368.323
	Goiás	73.522
	Minas Gerais	64.853
Total for the entire Centre-South Region		617.709
	Highlight	*Sugarcane (thousand tonnes)*
North-Northeast Region	Alagoas	16.382
	Pernambuco	11.394
	Paraíba	5.586
Total for the entire North-Northeast Region		49.115
Brazil		**666.824**

Source: ÚNICA (2015).

The importance of sugar cane in Brazilian agribusiness is indisputable and although Brazil stands out for all the technology already employed in the different stages of production, scientific research still has a lot to contribute to maximising the production process, from farming to industry (COSTA, 2005).

2.2 Botanical and morphological aspects of sugar cane

According to Caieiro et al (2010), sugar cane is a monocotyledonous, allogamous plant that reproduces sexually and can be multiplied asexually by vegetative propagation when grown commercially.

Classified according to Cronquist (1981), sugar cane belongs to:

- Division: Manoliophyta
- Class: Magnoliopsida
- Order: Graminales
- Family: Poaceae
- Genus: *Saccharum*
- Species: *Saccharum officinarum, Saccharum spontaneum, Saccharum sinensis, Saccharum barberi* and *Saccharum robustum.*

The commercial varieties of sugar cane are hybrids of the genus *Saccharum and* are thus called *Saccharum* spp. (CASTRO & KLUGE, 2001).

According to Bacchi (1983), sugar cane has the following morphological description:

1. The roots are fasciculated and can reach a depth of up to 4 metres;
2. Stalks are made up of nodes and internodes. At each node, there is a bud that is arranged alternately around the stem. The development of the buds results in shoots or primary stalks, from which tillers will later emerge. This process results in the formation of the sugarcane clump and the population of stalks that will be harvested;
3. The inflorescence and seeds of sugar cane, under certain conditions of photoperiod, temperature and humidity, flower by emitting a panicle or arrow;
4. The sugar cane flower is hermaphrodite, with only one ovule;
5. The pistils are terminated by reddish-purple stigmas, which give the panicle its characteristic feathery appearance;
6. The sugar cane seed is actually a caryopsis-like fruit with an elliptical shape, 1.5 mm long and 0.5 mm in cross-sectional diameter.

2.3 Sugar cane bagasse

According to Medeiros (1992), sugar cane bagasse *in natura* is defined as the residue of sugar cane stalks, the result of maximum extraction of the cellular content rich in soluble sugars. Therefore, sugar cane bagasse brings together coarse fragments of the cell wall and cell content not extracted during sugar cane milling, the main component of which is sugar not extracted during the milling process, approximately 2 to 3 per cent, and a high content of cell wall components (structural carbohydrates), around 70 to 85 per cent, of which cellulose is the main one (44 to 50 per cent), followed by hemicellulose (24 to 30 per cent) and lignin (10 to 20 per cent).

According to Alcarde (2009), sugar cane bagasse is the largest waste product of the Brazilian agro-industry. Approximately 280 kilos of bagasse per tonne of sugar cane are produced in the industrial process. Its main applications are: boiler fuel, pulp production and feed for confined cattle.

Furtado et al (2009), however, state that around 95% of all bagasse produced in Brazil is burnt in boilers to generate steam, producing bagasse ash as waste, the disposal of which in most cases does not follow appropriate practices and could pose a serious environmental problem.

2.4 Sugarcane bagasse x energy generation

The use of sugarcane bagasse to generate energy is due to Brazil's vast natural wealth, topography and favourable terrain for agricultural production, and the country has remained the leader in ethanol production since the early 1990s. This is mainly due to the vast amount of knowledge and technology about sugar cane, in terms of genetic improvement of the plant, combating pests, agricultural and harvesting techniques, the impact of the crop on the environment, and ethanol production technologies, including hydrolysis and fermentation (VASCONCELOS, 2003).

Macedo (2001) estimated that sugar mills could release between 30 and 50 per cent of the bagasse produced in sugar cane industrial processes for use in alternative energy generation routes.

According to the National Electric Energy Agency, there are currently 387 thermoelectric plants using sugarcane bagasse in operation in Brazil. This represents an installed capacity of 9,933,550 KWs, accounting for 25.9 per cent of the total energy produced by Brazilian thermoelectric plants (ANEEL, 2016).

Generating energy from biomass increases the prospect of

negotiating projects to commercialise carbon credits. Brazil is second only to India in the number of projects for commercialising carbon credits. The business that has most attracted foreign investors to Brazil is

country is the co-generation of energy from biomass. This segment already represents the majority of Brazilian projects in this market, and it is estimated, according to Table 02, that its potential for reducing emissions annually reaches 372,620,117 million tonnes of carbon in the country (MCTI, 2014).

Table 02 - Classification, by type, of Brazilian Clean Development Mechanism (CDM) projects registered up to 31 December 2014.

TYPE OF PROJECT	PROJECT NUMBERS	SHARE (%)
Hydroelectric	90	27,0
Biogas	64	19,2
Wind Power Plant	54	16,2
Landfill gas	50	15,0
Biomass Energy	41	12,3
Fossil Fuel Substitution	9	2,7
Methane avoided	8	2,4
Others	17	5,1

Source: MCTI (2015).

Martins (2009) considers the main economic advantage of producing electricity from sugar cane bagasse to be the fact that this process becomes a source of income when used under the rules of the Clean Development Mechanism (CDM) to commercialise carbon credits on stock exchanges.

According to Paoliello (2008), the energy utilisation of sugar cane bagasse is not a recent practice. Cogeneration is already widely used by mills and other industrial sectors, mainly to meet their own energy needs. There are even mills that make the surplus energy produced available to electricity distribution companies.

Various authors have reported on the importance of generating electricity using sugar cane waste. Martins (2009) considers the positive aspects to be:

a) Meeting the national need for electricity generation from new energy sources;

b) The production of electricity with totally clean technology, from a renewable source, which contributes to environmental preservation;

c) The production of electricity, especially during the rainy season, which coincides with the sugar cane harvest;

d) The inclusion of a new power generation agent, thus contributing to the consolidation of the new competitive market model;

e) A gain in competitiveness in the global sugar-alcohol sector, since a new product with stable revenue will be added from the better utilisation of a residual product;

f) The use of entirely national technology, preserving local jobs and relieving the country's balance of payments.

2.5 Poultry farming - Waste

The chicken industry in Brazil established itself as a modern segment in the 1970s, thanks to the agricultural policy of subsidised credit and the installation of slaughterhouses, as well as the links between national groups and foreign companies producing strains (RIZZI, 1993).

In 2004, Brazil became the world leader in chicken exports, overtaking the United States, which is the world's largest chicken producer. Brazilian exports and their increase were motivated by Avian Influenza, the outbreak of which, since the end of 2003, has damaged production and caused the sacrifice of more than 120 million birds in Asia (MARTINS, 2005).

According to the Brazilian Ministry of Agriculture, Livestock and Supply (MAPA, 2016), the country has become the world's third largest producer and leading exporter. Currently, the country's meat reaches 142 countries and the growth rate of chicken meat production, for example, is expected to reach 4.22 per cent per year. With exports expected to expand by 5.62 per cent per year, Brazil should continue to lead the world.

In Alagoas, according to the Brazilian Institute of Geography and Statistics, 7,028,078 head of chickens were registered in 2014, representing the total number of chickens in this last survey (IBGE, 2014).

As with other agricultural activities, poultry farming generates a large amount of waste (excreta, litter and dead birds) which, if well managed, could become not only an important source of income and added value to the activity, but also a sustainable production model that is increasingly becoming a market requirement. To this end, it is necessary to adopt a system for treating this waste in order to avoid possible contamination of the environment (ANGONESE etal., 2006).

The use of biodigesters to treat poultry waste allows systems to be integrated, as pointed out by Mahadevaswamy & Venkataraman (1986), who studied an integrated anaerobic biodigestion system to produce biogas from chicken waste and use the effluent to produce *Spirulinaplatensis* algae.

One sustainable use for chicken waste is to reuse it to produce biogas and biofertilisers, which is in line with the current need to develop new renewable energy sources (PALHARES, 2004; AIRES etal., 2009).

Studies by LUCAS JR. et al. (1993) on the use of additional inoculum in the performance of biodigesters operated with broiler waste showed an increase in potential and an anticipation of biogas production when this inoculum was used.

Salminen & Rintala (2002), carrying out a survey on the potential for methane generation from slaughterhouse and poultry farming waste, reported the importance of utilising this waste with potentials of 0.20 - 0.25 m^3 of CH_4 .kg "^{1} of poultry carcass; 0.10 - 0.15 m^3 of CH_4 .kg "^{1} of chicken litter; 0.05 m^3 of CH_4 .kg "^{1} of feather; 0.10 m^3 of CH_4 .kg "^{1} of blood, and 0.30 m^3 of CH_4 .kg^1 of viscera, feet and head.

According to Tessaro (2011), more recently there has been research into ways of utilising chicken waste to generate energy. It can be converted into energy through different processes, depending on the material and the type of energy required. Among these processes, fermentation is perhaps the most viable and, in some cases, direct combustion is an interesting alternative.

The production of biogas and biofertilisers in biodigesters using the anaerobic biodigestion process reduces costs on farms and/or industrial units and can help reduce the environmental impact of the production chain, as well as continuing to generate income from the sale of biofertilisers with high agronomic value (GIROTTO et al, 2003).

2.6 Biofuel

Biofuels are obtained through the transformation and fermentation of non-fossil biological sources such as vegetable oils, cereals and agro-industrial waste and can be used to replace or mix with conventional fuels (RODRIGUES, 2010).

Knight (2007) states that Brazil obtains three times more energy from biomass than the average country and five times more than most European countries. Considering that one of the most abundant raw materials in Brazil for biofuel production is sugar cane, and that the plant has a large biomass capacity, the Brazilian scenario is encouraging.

In the case of Brazil, it is estimated that the surplus sugar cane bagasse, if it were used to produce biofuel, would make it possible to double the country's production of this fuel without increasing plantation areas (BETANCUR, 2005; PEREIRA JR" 2006).

Dias (2011) observes that the full utilisation of sugar cane for biofuel production, through the pre-treatment and hydrolysis of surplus bagasse and sugar cane straw, is proving to be a good solution to problems such as competition for the use of land for food production and the expansion of sugar cane production through deforestation.

2.7 Biogas

Biogas comes from the activity of microorganisms (fermentation) and is made up of a mixture of different gases, including methane, carbon dioxide, hydrogen and sulphur dioxide. Biogas is flammable due to methane, a gas that is lighter than air, colourless and odourless. What causes the odour in biogas is sulphur dioxide which, even in small quantities, is perceptible to the sense of smell and quite corrosive (OLIVEIRA, 2004).

Different organic materials can be used to produce biogas. Waste from agriculture and agro-industry, from the most diverse compounds to animal excrement and faeces and sewage. The development of bacteria results in the breakdown of organic compost, which is transformed into biogas and biofertilisers free of harmful bacteria and zoonoses (MURARO, 2006).

Different chemical processes are known for producing biogas, but the fermentation process is still the most widely used because it is more economical. This advantage is mainly due to the large number of cheap natural raw materials (mainly sugar and starch) that can be used (SCHMIDELL et aL, 2001).

For the fermentation of organic matter to take place, the microorganisms need a favourable environment for their growth and development with an absence of toxic chemical compounds (soap, detergent); adequate temperature; presence of organic matter (waste); absence of air. Thus, if there is any interference in these factors, it can lead to a reduction in biogas production (SEIXAS; MARCHETTI, 1981).

Anaerobic biodigestion takes place in closed chambers, providing the environment with the necessary conditions for microorganisms to digest the organic matter present (TARRENTO and MARTINEZ, 2006).

It takes place in four stages: hydrolysis, acidogenesis, acetogenesis and methanogenesis (Figure 01), through the action of specific microbial populations (STEIL, 2001).

In hydrolysis, organic matter is transformed by enzymes excreted by fermentative bacteria into sugars, amino acids and peptides (VAN HAANDEL and LETTINGA, 1994).

Acidogenesis is of great importance because it is when dissolved oxygen is removed. At this stage, the products of hydrolysis are absorbed by the fermentative bacteria and excreted in the form of simple organic substances such as volatile fatty acids, alcohols, lactic acid and mineral compounds such as $CO2$, H_2 , NH_3 , $H_2 S$, etc. (VAN HAANDEL and LETTINGA, 1994).

In acetogenesis, acetic and propionic acids are formed, with a large generation of hydrogen and a lowering of the pH (CHERNICHARO, 1997).

In methanogenesis, methane is formed by acetoclastic methanogenic bacteria (acetate-utilising) and hydrogenotrophic methanogenic bacteria (hydrogen-utilising) (STAMS, 1994).

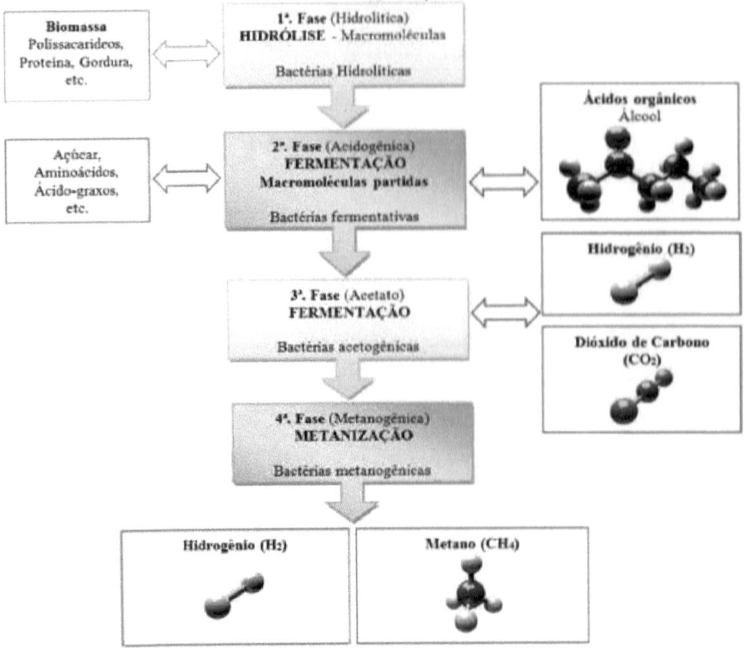

Figure 01 - Biogas formation processes.

Source: LETTINGA, HULSHOF; ZEEMAN.1996.

The process of generating biogas, although it normally occurs in nature, can be controlled through biodigesters built for this purpose (Figure 02).

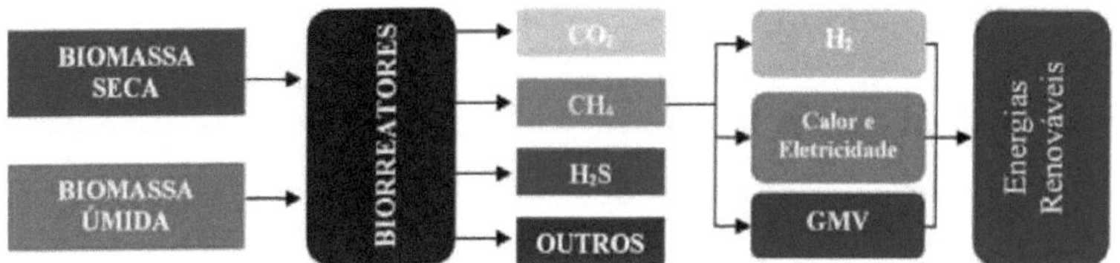

Figure 02 - Renewable energy generation scheme.

Source: VATTAMPARAMBIL (2012).

The most widespread biodigester models in the world were developed and perfected in China and India. In the 1950s and 1960s, at the height of the Cold War, communist China opted for a policy of energy decentralisation, with the aim of making small towns and agricultural centres self-sufficient, while the Indians, plagued by poverty and lacking self-sufficiency in oil, were forced to use their knowledge to make up for the shortage of hunger and energy (BARREIRA, 2011).

In Brazil, research into the use of biodigesters gained momentum in the 1980s and was mainly carried out in the southern region, where large pig, poultry and cattle farmers are concentrated. However, it was in the northeast that researchers became interested in using the biomass generated on small and medium-sized farms, due to the warm climate (average annual temperatures above 25°C), which favours the development and activity of anaerobic microorganisms (SOUZA;I_AGE FILHO, 2014).

CHAPTER 3

METHODOLOGY

3.1 Place of the experiment

The experiment was conducted at the Environmental Sanitation Laboratory of the Technology Centre (CTEC) of the Federal University of Alagoas (UFAL), in the municipality of Maceió, in 2016.

3.2 Collection and characterisation of sugarcane bagasse and chicken droppings

The sugar cane bagasse was collected at the Utinga Leão Mill, located in the municipality of Rio Largo/AL. The material consisted of approximately 5kg of sugar cane bagasse, milled shortly before collection through the industrial milling process for ethanol and sugar production. The size of the particles was between 4 and 5 cm and they were immediately stored in a plastic container to be taken to the experimental food analysis laboratory at CECA/UFAL.

The chicken droppings came from the Poultry Sector of the Agricultural Sciences Centre of the Federal University of Alagoas/UFAL, located in the municipality of Rio Largo/AL. The chickens were up to 10 days old and were fed corn and soya-based feed. Excreta was collected twice a day for five days and stored in a refrigerator for immediate freezing.

3.3 Setting up the experiment

A battery of 6 experimental batch-type biodigesters (reactors) with a capacity of 600 mL was used for the research, with the reaction volume corresponding to 300 mL of substrate and 300 mL of headspace (Figure 3).

Anaerobiosis was established by replacing all the atmospheric air in the reactors' headspace with nitrogen bubbling.

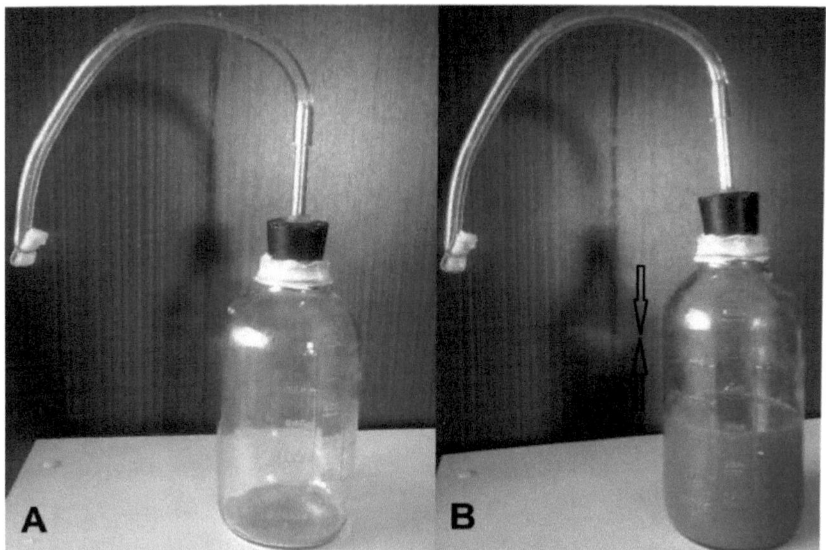

Figure 03 - Reactor (A) and Headspace (B).
Source: Author (2016).

The reactors were hermetically sealed using a rubberised stopper containing a metal rod to which a silicone hose was attached. The end of the hose was insulated with siliconised glue and bent to prevent the gas produced from escaping.

The reactors were closed immediately after the material to be fermented was packaged and only opened at the end of the experiment.

3.4 Experimental treatments

Two treatments with three replicates were tested. Sugar cane bagasse was used as the substrate and chicken droppings as the inoculum.

Treatment 01 consisted of 20 grams of ground and dehydrated *fresh* sugar cane bagasse, 10 grams of dehydrated and ground chicken droppings and distilled water to complete 300 ml, 50 per cent of the reactor volume.

Treatment 02 consisted of 20 grams of *freshly* ground and dehydrated sugar cane bagasse, 10 grams of dehydrated and ground chicken droppings and a 4% NaOH solution of distilled water, up to 300 ml, corresponding to 50% of the reactor volume (Table 03).

Table 03 - Description of the experimental treatments

Treatments	Inoculo	Substrate	Distilled water
T1R1	10g	20g	Distilled water up to 300 ml from the reactor
T1R2	10g	20g	Distilled water up to 300 ml from the reactor
T1R3	10g	20g	Distilled water up to 300 ml from the reactor
T2R1	10g	20g	4% NaOH solution up to 300 ml of the reactor
T2R2	10g	20g	4% NaOH solution up to 300 ml of the reactor
T2R3	10g	20g	4% NaOH solution up to 300 ml of the reactor

Source: Author (2015).

3.5 Biogas collection

The biogas was collected using a "gastight" syringe with a lock, which passed through the silicone hose implanted in the nozzles of the containers, and was collected three days a week - Monday, Wednesday and Friday (Figure 04). After collection, the hole used to aspirate the gas was immediately closed with the silicone glue.

Figure 04 - Gastight syringe used to collect biogas.

Source: Author (2016).

3.6 Characterisation of the material used

The experiment ran for 60 days, during which time the CH values$_4$ (mmol.day^{-1}) produced by each batch reactor were measured.

The temperature observed throughout the experiment was an average of 26 °C and was controlled against drastic and sudden changes by placing the reactors in a Styrofoam box.

To characterise the materials used as substrate and inoculum, the dry matter (DM), ash (APHA, 1992), crude protein (CP) and nitrogen (N) contents were analysed using the Kjeldahl method, according to Silva (2003).

To characterise the initial and final sludge, in addition to the above, total solids (TS), total volatile solids (TVS), total fixed solids (TFS), pH, chemical oxygen demand (COD) and COD efficiency (APHA, 2002) were observed.

a) **Dry matter:** Determination of the weight loss of the product when heated to constant weight. The sample was heated directly to a temperature of 105°C.

b) Ash: Residue obtained by incineration at temperatures of 550° to 570°C.

c) Crude protein and nitrogen: The analysis was carried out using the Kjeldahl method, where the percentage of nitrogen obtained is multiplied by 6.25 and then expressed as Crude Protein (CP). This analysis is based on the fact that all proteins contain 16% nitrogen, and that all the nitrogen in the food is in protein form.

d) Total Solids (ST): The initial weight (P_o) of the sample was measured as soon as it was removed from the Mufla at a temperature of 550°C, after 60 minutes. The sample was cooled and its weight measured. Next, 50 mL of the raw sample, measured in a 100 mL beaker, was transferred to the capsule and placed in an oven at 103° - 105°C for approximately 24 hours. Once out of the oven, they were placed in a desiccator until they cooled down and the capsule was weighed again, thus obtaining the dry weight (Pi), which was determined using Equation 01:

$$\mathbf{ST\ (mg/l) = (P_1 - P_0).\text{-}10^6}\ {}_{/VAsample}$$

Where:

P_1 : Dry weight

P_0 : Initial weight

V: Volume

e) Total Volatile Solids (VTS): After weighing, the sample was removed from the oven and subjected to calcination in a muffle furnace at 550°C for 60 minutes, then removed and cooled in a desiccator, weighed on a precision scale, and the final weight was obtained (P_2). Calculations expressed by Equation 02:

$$STV\ (mg/l) = \mathbf{(P_1 - P_2).10^6}\ {}_{/VAsample}$$

Where:

P_1 : Dry weight

P_2 : Final weight

V: Volume

f) Total Fixed Solids (TFS): Portion of the residue after calcination at 550°C for 1 hour (APHA, 1992). Expressed by Equation 03:

$$STF\ (mg/l) = \mathbf{(P_2 - P_0).10^6}_{/VAsample}$$

Where:

P_2 : Final weight

P_0 : Initial weight

V: Volume

g) pH - Hydrogen Potential: A potentiometer with a calomel electrode was used to measure the pH of the initial and final samples. No pH correction was made throughout the analysis process.

h) COD - Chemical Oxygen Demand: Measurements were made using a HACH spectrophotometer, model DR-2500, with a wavelength of 620 nm, after digestion of the organic matter at 150°C for 120 minutes.

3.7 - Analysing the biogas produced - CH_4

The methodology applied by Maintinguer et al. (2008) was used to determine the production and composition of the biogas produced using a *Shimadzu GC-2010-Plus®* gas chromatograph. The device has a thermal conductivity detector and a 30 m long *Supelco Carboxen 1010 Plot®* column with an internal diameter of 0.53 mm. The gases produced in the reactors were collected using a "gastight" syringe (1 ml) with a lock, and then applied to the chromatograph (Figure 05).

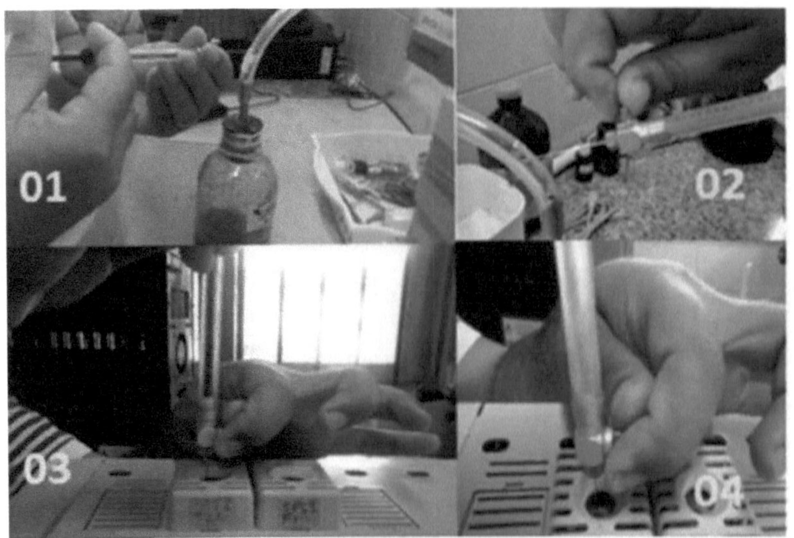

Figure 05 - Biogas collection (01), locking (02), unlocking (03) and application (04).

Source: Author (2016).

Collections began on the fifth day after the reactors were closed for good. An attempt was made to maintain regular readings three times a week, thus totalling 54 days of reactor operation. The Boltzmann Linear Equation was used to adjust the data collected:

$$Y = [(A_1 - A_2) / ((1 + e^{(x-x0)})/d_x)] + d_x$$

Where:

A< Initial value = 0

A_2 : Final value = 1

$_{x0}$: Centre = 0

d_x : Time constant = 1

To determine the highest methanogenic production in the reactors, the Gompertz equation was used, applied using *Software-origin* 11:

$$Y = a^* \exp(-\exp(-k^*(x-xc)))$$

CHAPTER 4

RESULTS AND DISCUSSION

As for the characterisation of the sugarcane bagasse and chicken droppings used to supply the reactors, Table 04 shows that the values for dry matter, crude protein and nitrogen were not very different when comparing the two residues. However, there was only a more marked difference in the ash values, 20.08% and 7.36% for sugar cane bagasse and chicken droppings, respectively.

Table 04 - Characterisation of the material used

	Chicken droppings	Sugar cane bagasse
Dry matter (DM)	77,36 %	95,36 %
GREY (MS)	20,08 %	7,36 %
Crude Protein (DM)	9,68 %	9,38 %
N (MS)	1,55%	1,50%

Source: Author (2016).

Different values for the bromatological analyses of sugar cane bagasse have been recorded by researchers. This is due to differences between sugar cane varieties, cutting age and other factors that interfere with these results.

Carvalho et al (2009) recorded dry matter values for sugar cane bagasse at 40 per cent. Similar results were observed by Carvalho et al (2006), who recorded 40.11 per cent. Cândido et al (1999), however, found values of 69.6% for the dry matter present in sugar cane bagasse.

When analysing the ash content of sugarcane bagasse, Fredericci et al (2012) state that the literature indicates various levels of ash obtained from burning sugarcane bagasse, ranging from 2.4% to 10% by weight. The 7.36 per cent observed in this study therefore corroborates Fredericci et al (2012).

Carvalho et al (2006) observed crude protein contents of 2.32% when they characterised the bromatological chemical composition of sugar cane bagasse.

Cândido et al (1999) stated that the minimum level of crude protein considered for the best performance of fermentation activities is 7 per cent.

The observed nitrogen contents were 1.55% and 1.50% for chicken excreta and sugar cane bagasse respectively. Due to the similar nitrogen levels recorded between the residues, it is suggested that chicken excreta may not have been necessary as an N-providing inoculum. However, due to its high content of microorganisms necessary for the fermentation process, it continued to be inoculated in the formation of the reaction slurries.

Analysing the initial and final sludge collected from both the reactors containing *raw* sugarcane bagasse and the reactors containing treated sugarcane bagasse, the following values were observed (Table 05).

Table 05 - Characterisation of the initial and final *lots*

	Beginning fudge		Final round	
	Raw sugarcane bagasse	Treated sugarcane bagasse	*Raw sugarcane bagasse*	Treated sugarcane bagasse
Dry matter	87,52 %	84,31 %	88,16%	86,94 %
GREY	6,05 %	47,42 %	8,76 %	48,04 %
Crude Protein	6,71 %	4,89 %	7,11 %	5,93 %
N	1,07%	0,78 %	1,14%	0,95 %
ST (mg/l)	42.932,22	96.712,30	37.231,70	63.871,88
STV (mg/l)	24.902,97	46.711,26	34.404,52	34.529,93
STF (mg/l)	18.029,25	50.001,04	2.827,19	29.341,95
Ph	5,09	13,05	3,60	13,08
COD (mg/l)	70.148,88	77.524,32	26.818,17	23.437,76
Efficiency COD			61,79%	69,41 %

Source: Author (2016).

A decrease in total, volatile and fixed solids was observed. This behaviour was identical to that of other experiments conducted with different inputs.

Santos (2016), in experiments with anaerobic reactors supplied with *citrus* waste combined with pig manure, also observed a decrease in the values for the three solids indices.

Steil (2001), analysing the use of inoculums in anaerobic digestion of poultry and pig waste, found that there was also a reduction in the total, volatile and fixed solids determined in his experiment.

When analysing the pH content of the initial and final sludge present in the anaerobic reactors in this experiment, it was possible to see that there was a decrease in the values for the treatments made up of *raw* sugarcane bagasse. However, for the reactors supplied with sugarcane bagasse treated with a 4% NaOH solution, this index increased from 5.09 to 3.60 for the first treatment and from 13.05 to 13.08 for the second treatment. This reaction of keeping the pH practically stable for the treated sugar cane bagasse was probably due to the fact that the sugar cane bagasse was treated inside the reactor at the beginning of its assembly. This allowed the reaction environment to be directly influenced by the 4% NaOH solution.

According to Chermicharo (1997) and Van Haandel & Lettinga (1994), the ideal pH for methane formation should vary between 6.0 and 8.0. However, they state that values below 6.0 and above 8.3 should be avoided as they inhibit the bacteria that form this gas.

The methanogenic production (mmol.day^{-1}) observed over 54 days, from the closure to the opening of the reactors, can be seen in Figures (06 and 07):

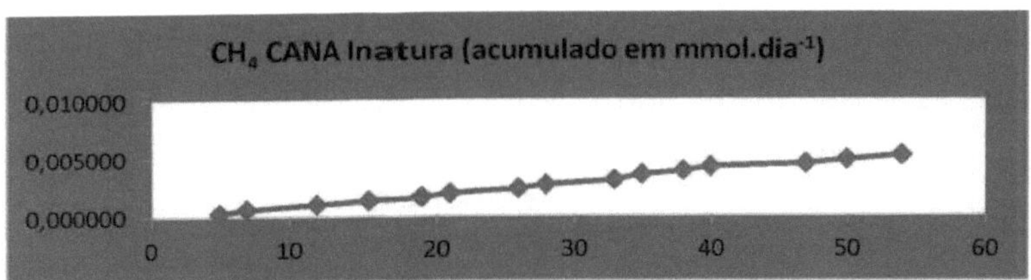

Figure 06 - Accumulated CH_4 in mmol.day^{-1} , produced by sugar cane bagasse *in natura.*

Source: Author (2016)

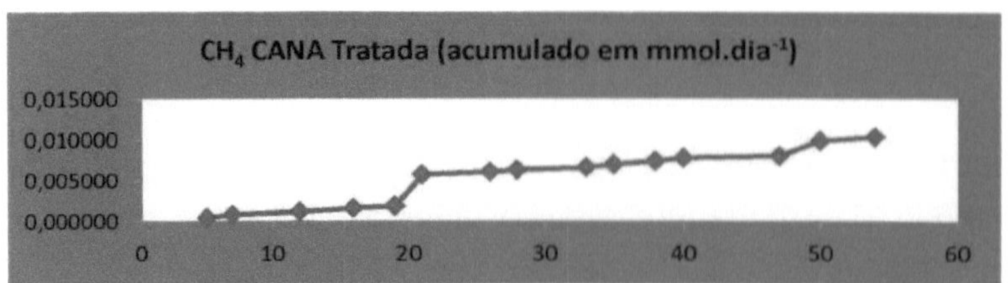

Figure 07 - Accumulated CH_4 in $mmol.day^{-1}$, produced by treated sugarcane bagasse.

Source: Author (2016).

A comparison of the data represented in the two figures shows that the reactors containing sugarcane bagasse treated with sodium hydroxide (NaOH) obtained a greater accumulation of methane gas (0.000494 mmol) over the 54 days analysed, when compared to the reactors containing *raw* sugarcane bagasse (0.000345 mmol). This difference can be attributed to the action of sodium hydroxide (NaOH) on the cellulose and lignin molecule during the reactors' gas formation process, acting to hydrolyse the ligno-cellulose compound.

Lignin is a hydrophobic, non-polysaccharide biomolecule that interacts with cellulose and hemicellulose. It is responsible for almost all the rigidity of the cell wall (HENRIKSSON, 2009).

Siqueira (2015), in research on sugarcane bagasse, which was pre-treated with NaOH in a 5% alkaline solution to assess the hydrolysis of lignin and cellulose, observed that the action of the solution used achieved a high level of solubilisation of the lignin, leading to better adsorption.

Reactions in an alkaline environment favour solubilisation as they result in the depolymerisation of lignin. In acidic conditions, the condensation of lignin increases its molar mass and decreases its solubility (FENGEL; WEGENER, 1983; GELLERSTEDT, 2009).

After applying the Gompertz equation, it was possible to determine the day on which the reactors obtained the highest methanogenic production rate. The results are shown in Figures 08 and 09.

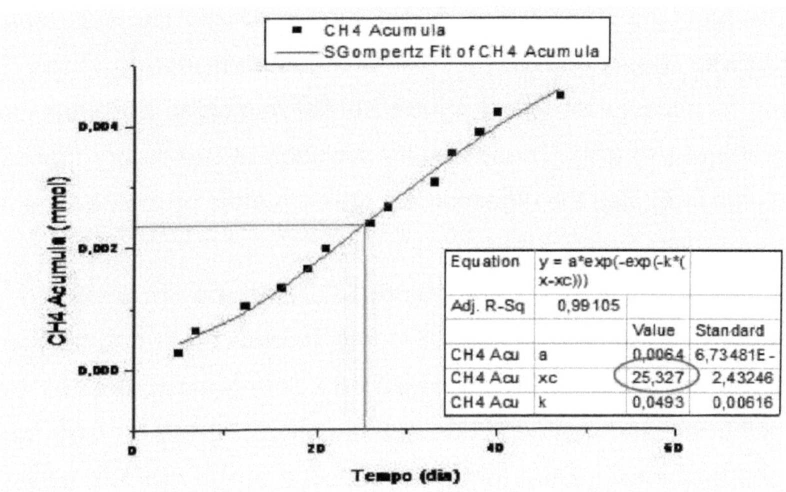

Figure 08 - Methanogenic enhancement of CH_4 in reactors with *raw* sugarcane bagasse.

Source: Author (2016).

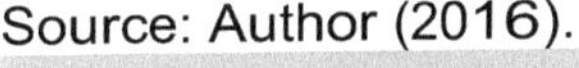

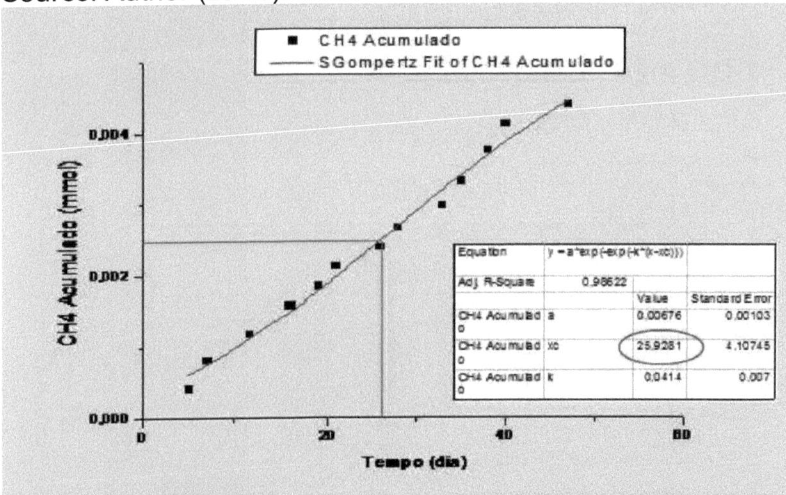

Figure 09 - Methanogenic enhancement of CH_4 in reactors with treated sugarcane bagasse.

Source: Author (2016).

Figures 8 and 9 show that the approximate date for the highest production efficiency of CH_4 is the twenty-fifth day (25.327) for *raw* sugarcane bagasse and the twenty-sixth day (25.9281) for treated sugarcane bagasse.

Comparing the days during which optimisation took place and the accumulation of methane gas, it was found that although the sugarcane bagasse treated with NaOH had an approximately equal period to obtain its best productivity, its accumulated productive potential was close to double that of the *raw* sugarcane bagasse. These results corroborate the theory that the breakdown of lignin facilitates the digestion of organic matter by methanogenic bacteria.

When the chemical oxygen demand (COD) values are analysed in the sludge of the two treatments studied (Figures 10 and 11), it can be seen that both treatment 1 and treatment 2 showed very close values, both for the initial sludge of the *raw* and treated sugarcane bagasse (70,148.88 mg.r and 77,524.32 mg.l , respectively) and for the final sludge of the *raw* and treated sugarcane bagasse (26,818.88 mg.r , respectively).148.88 mg.r^{-1} and 77,524.32 mg.l^{-1} , respectively), and for the final sludge of the *raw* and treated sugar cane (26,818.17 mg.l^{-1} and 23,437.76 mg.l^{-1} , respectively).

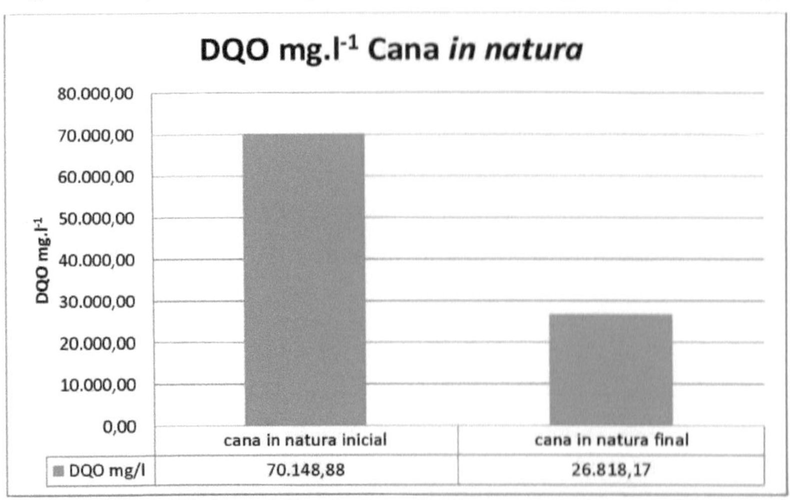

Figure 10 - Initial and final chemical oxygen demand in reactors with *raw* sugarcane bagasse.
Source: Author (2016).

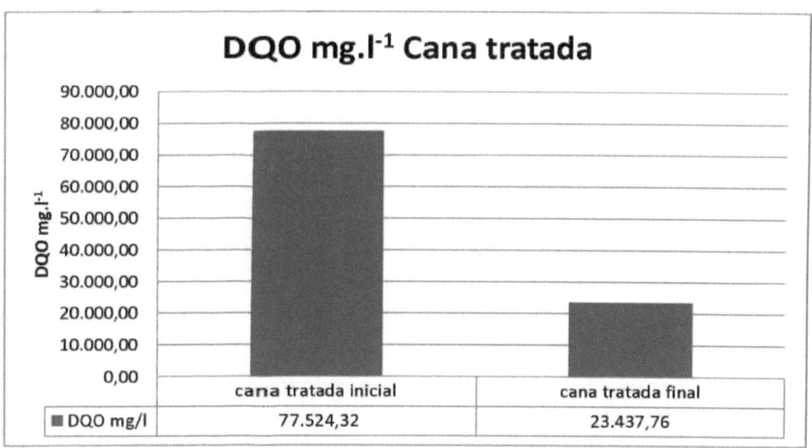

Figura 11 - Initial and final chemical oxygen demand in reactors with treated sugarcane bagasse.

Source: Author (2016).

Analysing the chemical oxygen demand efficiency in percentage, it was observed that treatment 2 with sugarcane bagasse treated with NaOH performed better than treatment 1 with *raw* sugarcane bagasse, recording values of 69.41% and 61.79% respectively (Figure 12).

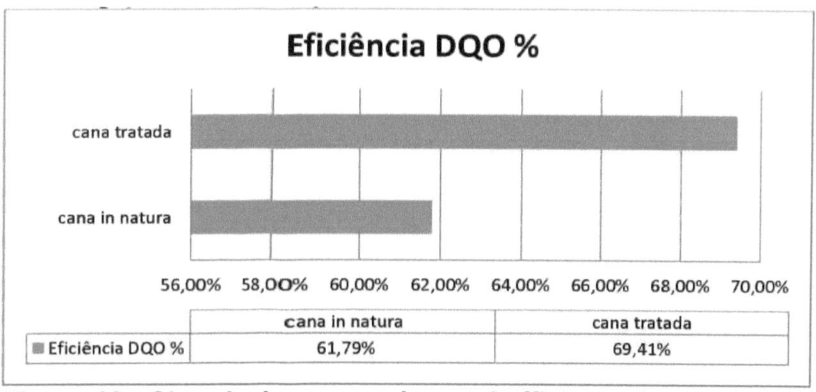

Figure 12 - Chemical oxygen demand efficiency in reactors with *raw and* treated sugarcane bagasse.

Source: Author (2016).

CHAPTER 5

CONCLUSIONS

The sugarcane bagasse treated with a 4% sodium hydroxide solution (NaOH) showed better results in CH4 production when compared to the results of the *raw* sugarcane bagasse, obtaining twice as much methane gas produced at similar times for maximum methanogenic production. Values of 0.010379 and 0.005185 were recorded during the 60 days of execution, in accumulated mmol, respectively.

REFERENCES

AIRES, A. M. **Anaerobic biodigestion of broiler litter with or without separation of the solid and liquid fractions.**2009. 134f. Dissertation (Master's degree in Zootechnics) - Faculty of Agricultural and Veterinary Sciences, Universidade Estadual Paulista, Jaboticabal, SP. 2009.

ALCARDE, A. R. **Sugarcane processing: other products.** EMBRAPA Information Agency, 2009. Available at: www.agencia.cnptia.embrapa.br. Accessed on: 32 September 2016.

ANGONESE, A.R. *et al.* Energy efficiency of a pig production system with waste treatment in a biodigester. **Revista Brasileira de Engenharia Agrícola e Ambiental,** Campina Grande, v.10, n.3, p.745-750, 2006.

APHA, AWWA, WPCF. **Standard methods for examination of water and wastewater.** 18. ed. Washington: APHA, 1992.

APHA, WPCF. **Standard methods for examination of water and wastewater.** 18. ed. Washington: APHA, 2002.

BACCHI, O. O. S. **Botany of sugar cane In: Orlando Filho, J. Nutrição e adubação da cana-de-açúcar no Brasil.** Piracicaba: IAA/PLANALSUCAR, 1983. p.25-37.

BARREIRA, Paulo. **Biodigesters: energy, fertility and sanitation for rural areas.** São Paulo: icon, 2011.

BETANCUR, G.J.V. **Advances in Hemicellulose Biotechnology for Ethanol Production by Pichiastipitis.** 2005. Master's dissertation. Federal University of Rio de Janeiro. Brazil. 2005.

CAIEIRO, J. T. et *al.* Physical purity and germination of sugarcane seeds (Caryopses) *(Saccharum* spp.). **Revista Brasileira de sementes,** Londrina, v. 32, n. 02, p. 140-145, Jun. 2010.

CÂNDIDO, M. J. D. et *al.* Evaluation of the Nutritional Value of Sugarcane Bagasse Ammonised with Urea. **Rev. bras. zootec.,** v.28, n.5, p.928-935, 1999.

CARVALHO, G. G. P. et *al.* Chemical composition and dry matter digestibility of sugarcane bagasse treated with calcium oxide. **Arq. Bras. Med. Vet. Zootec., v.61, n.6, p. 1346-1352, 2009**

CARVALHO, G. G. P. et *al.* Nutritional value of sugarcane bagasse ammonised with four doses of urea. **Pesq. agropec. bras.,** Brasília, v.41, n. 1, p.125-132, jan. 2006

CASTRO, P. R. C.; KLUGE, R. A.. **Ecophysiology of stratigraphic crops: sugar cane, rubber, coconut, oil palm and olive.** Cosmópolis: Stoller do

Brazil, 2001, p. 13-45.

CHERNICHARO, C. A. L. **Anaerobic reactors: principles of the biological treatment of wastewater.** Belo Horizonte: Department of Sanitary and Environmental Engineering (DESA/UFMG), 1997. 246p.

COSTA, M. C. G. **Distribution and root growth in sugarcane ratoons: two cultivars in soils with different characteristics.** 2005. 88 f. Thesis (Doctorate in Agronomy) Escola Superior de Agricultura Luiz de Queiroz, Universidade de São Paulo, Piracicaba, 2005.

CRONQUIST, A. **An integrated system of classification of flowering plants.** New York: Columbia University Press, 1981. 126p.

DIAS, M.O.S. et al. Second generation ethanol in Brazil: Can it compete with electricity production? **Bioresource Technology,** v. 102, p.8964-8971, 13 Jul. 2011.

FAHL, J.I. *et al.* Polyoses Hemicelluloses. In: **WOOD: chemistry, ultrastructure,** reactions.Berlin: Walterde Gruyter, 1983.

FAHL, J.I. et *al.* **Agricultural instructions for the main economic crops**, Bulletin 200, ed. 6, IAC, 396p. 1998.

FREDERICCI, C. et *al.* Characterisation of sugarcane bagasse ash as a raw material for ceramic production. CONGRESSO BRASILEIRO DE CERÂMICA, 56, and CONGRESSO LATINO-AMERICANO DE CERÂMICA,

1, and BRAZILIAN SYMPOSIUM ON GLASS AND RELATED MATERIALS, 9. 03 to 06 June 2012, Curitiba, **Anais...** Curitiba, 2012.

FURTADO, A. D. et *al.* **By-products of the sugar-alcohol industry and their use** - Module VI Specialisation Course in Management in the Sugar-Alcohol Industry, 1. ed. Campina Grande - PB: Centre for Technology and Natural Resources - Agricultural Engineering Academic Unit of the Federal University of Campina Grande - Cuiabá - MT: Faculty of Agronomy and Veterinary Medicine - Department of Soils. 2009.

GELLERSTEDT, G. Chemistry of Chemical pulping. In: EK, M.; GELLERSTEDT, G.; HENRIKSSON, G.. **Pulping chemistry and technology.** Berlin: Walterde Gruyter, 2009.

GIROTTO, A. F., ÁVILA, V. S. **Poultry litter: economic analysis of alternative materials.** Concórdia: Embrapa Swine and Poultry, 2003. 4p. Embrapa Swine and Poultry. Technical Communication, 326.

HENRIKSSON, G. Lignin. In: EK, M.; GELLERSTEDT, G.; HENRIKSSON, G.. **Wood chemistry and biotechnology.** Berlin: Walter de Gruyter, 2009.

IBGE **- Brazilian Institute of Geography and Statistics -** Systematic Survey of Agricultural Production, 2016. Available at: http://www.ibge.gov.br/home/estatistica/indicadores/agropecuaria/lspa/lspa_2 01608 _5.shtm. Accessed on: 19 June 2016.

ISRAEL, C.M. **Utilisation of palm processing waste for the production of hydrolytic enzymes by fungi of the genus Polyporus.** 2005. 136 p. Dissertation (Master's in Environmental Engineering) - Regional University of Blumenau, Blumenau, 2005.

Knight, P. Energy crops have begun a revolution for agribusiness in Brazil. World **Ethanol & Biofuels Report, F. O. Licht,** v.05, n.12, p. 243-248, 2007.

LETTINGA, G. Anaerobic treatments of very low strength and cold industrial and domestic wastewater. In: VINHAS, M; SOUBES, M; BORZACCONI, L; MUXI, L. (eds). TALLER - LATIN AMERICAN SEMINAR - Anaerobic treatment of wastewater, 3, 1994 Montividéo, **Anales...** Montividéo, Uruguay, 1994. P. 155168.

LETTINGA, G; HULSHOF, P. W; ZEEMAN, G. Biological Wastewater Treatment. Part I: **Anaerobic Wastewater treatment.** Lecture Notes. Wageningen Agricultural University, Ed. January, 1996.

LIMA, G. A. **Sugarcane culture.** Fortaleza: Editora Fortaleza. 1984. 159p.

LUCAS JÚNIOR, J. et *al.* Evaluation of the use of inoculum in the performance of biodigesters fuelled with broiler manure and wood shavings. In: BRAZILIAN CONGRESS OF AGRICULTURAL ENGINEERING, 22. 1993, Ilhéus. **Proceedings...** Ilhéus: SBEA/CEPLAC, 1993. v. 2, p. 915-30.

MACEDO, I. C. **Electricity generation from biomass in Brazil: current situation, opportunities and development,** *Report for MCT,* Brasília, 2001.

MAHADEVASWAMY, M" VENKATARAMAN, L.V. Bioconversion of poultry dorppings for biogas and algal production. **Agricultural Wastes,** v.18, n.2, p.93-101, 1986.

MAINTINGUER, S. et *al.* **IFermentative** Hydrogen Production by Microbial Consortium. **International Journal of Hydrogen Energy,** v. 33, p. 4309 - 4317, 2008.

MAPA - **Ministry of Agriculture, Livestock and Supply** - Sugarcane. Available at: http://www.agricultura.gov.br/vegetal/culturas/cana-de-acucar. Accessed on: 01 August 2016.

MARTINS, R. **Energy produced from sugarcane bagasse is economically viable.** USP Agency: 2009. Available at: http://www.essentiaeditora.iff.edu.br/index.php/BolsistaDeValor/article/viewFile/2433/ 1321. Accessed on: 18/08/2016.

MARTINS, S. S. Poultry farming: situations and prospects in May 2005. **Informações Económicas,** v.35, n.7, São Paulo: IEA, 2005, pp. 57-59.

MCT, Ministry of Science and Technology. **Plants take advantage of co-generation and profit from the carbon credit market,** 2006. Available at: http://www.abepro.org.br/biblioteca/enegep2010 . Accessed on 11 July 2016.

MEDEIROS, R.S. **Effect of replacing self-hydrolysed sugar cane bagasse with sorghum on rumen fermentation in cattle, in vitro digestibility in sheep and animal performance in finishing cattle.** 1992. 104p. Dissertation (Master in Animal and Pasture Science) - Escola Superior de Agricultura "Luiz de Queiroz", Universidade de São Paulo, Piracicaba, 1992.

MURARO, W. **Evaluation of ice engine operation with low calorific value gas from rice husk gasification.** Master's dissertation - Postgraduate programme at the School of Mechanical Engineering. Campinas, 2006.

OLIVEIRA, P. A. V. de. **Technologies for waste management in pig production: manual of good practices.** Concórdia: Embrapa Swine and Poultry, 2004 (National Environment Programme - PNMA II).

PALHARES, J. C. P. **Use of chicken litter in biogas production.** Concórdia, SC: Embrapa Swine and Poultry, 12p. (Technical Circular, 41).

2004.

PAOLIELLO, J. M. M. **Environmental Aspects and Energy Potential in the Utilisation of Waste from the Sugar and Alcohol Industry.** 2008. Master's dissertation. Faculty of Engineering, São Paulo State University, Bauru, 2008.

PELIZER, L.H.; PONTIERI, M.H.; MORAES, I.O. Utilisation of Agroindustrial Waste in Biotechnological Processes as a Perspective for Reducing Environmental Impact. **Journal of Technology Management & Innovation,** V.2, p. 118-127, 2007.

PEREIRA Jr., N. Lecture II: Lignocellulosic Biotechnology and the Associated Biorefinery Context. **Brazilian Congress of Chemical Engineering (COBEQ),** São Paulo, 2006. **16. Proceedings...** 24-27 September 2006.

RIZZI, A. T. **Technological Change and Restructuring in the Food Industry: the case of the chicken industry in Brazil.** 1993. Thesis (Doctorate in Economics). Campinas: Unicamp, 1993.

RODRIGUES, L. D. **Sugarcane as a raw material for biofuel production: environmental impacts and agroecological zoning as a mitigation tool.** Federal University of Juiz de Fora, Minas Gerais, 2010.

SALMINEN, E.A.; RINTALA, J.A. Semi-continuous anaerobic digestion of solid poultry products.

slanghterhouse waste: effect of hydraulic retention time and loading.Water **Research,** London, v.36, p.3.175-3.182, 2002.

SANTOS, I. V. V. S. **Anaerobic biodigestion of citrus agroindustry waste in consortium with pig manure.2016.** Master's dissertation. Master's Degree in Biomass Energy. Federal University of Alagoas, 2016.

SCHMIDELL, W.; LIMA, U.S.; AQUARIBE, E. & BORZANI, W. **Biotecnologia Industrial: Engenharia Bioquímica**, v. 2. São Paulo: Editora Edgard Blucher, 1 edição. 541 p. 2001.

SEIXAS, J; MARCHETTI, D. A. B.. **Construction and operation of biodigesters.** Brasília: Embrapa Swine and Poultry, 1981. 60 p.

SINDAÇÚCAR - Sugar and Alcohol Industry Union - **General Production Bulletin, 2016.** Available at: http://www.sindacucar-al.com.br/wp-content/uploads/2016/09/N24BoletimgeraldeproducaoART20152016.pdf. Accessed on: 01 June 2016.

SIQUEIRA, G. A. **Effect of lignin from pre-treated sugarcane bagasse on the enzymatic hydrolysis of cellulose.** 2015. 139p. Thesis (Doctorate in

Science) - Lorena School of Engineering, University of São Paulo, Lorena - SP, 2015.

SOUZA, M. L. B.; LAGE FILHO, F. A. Use of anaerobic biodigesters to utilise animal waste and control environmental pollution. **Oswaldo Cruz Academic Journal,** year 1, n.3 July-September 2014.

STAMS, A. J. M. Metabolic interactions between aerobic bacteria in methanogenic environments. **Antonie van Leeuwenhoek,** v.66, p. 271-294, 1994.

STEIL, L. **Evaluation of the use of inoculums in the anaerobic biodigestion of poultry, broiler and pig waste.** 2001. 109f. Dissertation (Master's Degree in Biotechnology) - Chemistry Institute, Paulista State University, Araraquara, 2001.

TARRENTO, G. E.; MARTINEZ, J. C. Analysis of the implementation of biodigesters in small rural properties, within the context of clean production. In: SIMPEP, 13. Bauru, Anais...Bauru, SP, 2006.

TESSARO, A. A. **Energy potential of poultry litter produced in the south-western region of Paraná used as a substrate for biogas production.** 2011. Dissertation (Master's in Engineering). Institute of Technology for Development in partnership with the Engineering Institute of Paraná, Paraná: Curitiba, 2011.

UNICA - **Union of the Sugarcane Industry** - History of production and milling, 2015. Available at: http://www.unicadata.com.br/historico-de-producao- e-moagem.php?idMn=31&tipoHistorico=2. Accessed on: 20 June 2016.

VAN HAANDEL, A. C.; LETTINGA, G. **Anaerobic sewage treatment: a manual**

for hot climate regions. Campina Grande: Epgraf, 1994. 21 Op.

VASCONCELOS, Y. There's not even any bagasse left. 2003. **Pesquisa Fapesp magazine.** São Paulo, ed. 77. Available at: www.revistapesquisa.fapesp.br/2013. Accessed on: 11 July 2016.

VATTAMPARAMBIL, S. R. **Anaerobic Microbial Hydrolysis of Agriculture Waste for Biogas Production.** International Conference on Emerging Frontiers in Technology for Rural Area (EFITRA). Proceedings published in International Journal of Computer Applications - 2012.

Printed by Books on Demand GmbH, Norderstedt / Germany